Degradation Mechanisms and
Inspection Methods
Of Process Equipment
In Oil / Gas and Power Industries

Series Books on Inspection

BOOK-No.2

INSPECTION OF STEAM BOILERS AND HEAT EXCHANGERS

Authors: Fereydoun Majdnia
            Evgeny Sergeev

ISBN: 9781071149478

Majdnia, Fereydoun
Sergeev, Evgeny

Table of Contents

PREFACE

The authors are experienced inspection engineers with years of work in oil and gas industry with exposure to various types of the projects, operations and have worked in the field of inspection, training of inspectors, establishing quality system program for projects and inspection program for the operational plants.

In line with the training of the inspectors, students and those interested in inspection of plant equipment, the authors decided to prepare book series on inspection of the equipment to be used by interested persons. The unique approach is, in addition to explaining the inspection methods the common degradations and problems of the equipment are also described.

We see it more useful to design small books on the inspection of various types of equipment rather than a single and bulky book to cover all items. The pocket type and handy books are easily accessible and are mobile to be used by students, inspectors and technicians with affordable price. These are not made for the shelves but for the desktop and pocket.

We have tried to cover the type of the equipment that are used widely in various processing plant such as crude oil refineries, petrochemical plants, boosting pump stations, tank farms and loading facilities, oil and gas processing plants as well as power generation and utility plants.

It worth to note that we have avoided to concentrate on inspection check lists and reporting formats as those can easily be prepared and designed by individuals and available in each company.

This series of books covered so far are the following titles:

- Book-1 Heat Treatment of the weldments

- Book-2 Inspection of Steam Boilers and Heat Exchangers

INTRODUCTION

The purpose of this series of book is to discuss and explain the inspection of the Processing equipment used in Oil and Gas production facilities and assist new inspectors and technicians in developing the basic skills. The intention is not to talk about the inspection of equipment during the fabrication and assembly at shop which requires different approach and technique, but it is rather to concentrate on the in-service inspection.

The reason for the inspection of a process equipment which has been in service for a period of times, is to gather data on the condition of the equipment and parts so that it can be analyzed and a reasonable assessment is made of the equipment's mechanical integrity for continued service or scope of the repair, upgrading, alteration and or replacement as required.

This book is dedicated to steam boilers and heat exchangers which are used in oil and gas plants, power plants and food industries. It is of great importance to get familiarized with the common problems and degradation mechanisms in equipment especially the boilers and ensure they operate in safe condition as the failure of boilers has been catastrophic in the past 100 years.

The availability and reliability of the boilers and heat exchangers are of the prime concern in each operating unit. Plant management normally try all means to ensure these are achieved in safest and economical ways. Most of the companies now-a-days have adopted an asset integrity management system and follow the principals to achieve the goal by concentrating on the assets and their reliability.

The operating companies may have their in-house inspectors or use external inspectors to perform the inspection and monitoring to satisfy the integrity needs as well as regulatory requirements.

The inspectors must be trained adequately to be able to gain knowledge of codes and governing standards and be certified to perform the inspection. Each country has certain requirements on qualification and competence of the inspectors that must be followed by all involved in inspection.

In the absence of regulatory and jurisdiction requirements, each oil and gas producing company must ensure that the inspectors have adequate training and competence to perform the job and satisfy the requirements of insurance company and/or not to jeopardize the warranty of the equipment, if it is applicable.

It is advisable that the companies who possess and operate boilers and other pressure equipment use external certified inspectors such as API certified or NBBI commissioned or equivalent for inspection of the boilers in-order to have a more reliable survey and to train their own inspectors on the job.

Each operating company must have a set of written procedures for the inspection of boilers and other pressure equipment, safety requirements, environment protection, qualification and training requirements of personnel performing the inspection and/or examinations and qualification of those involved in the repair of the equipment.

On completion of the inspection, a review should be made by inspector/ engineer to verify the integrity of the equipment if degradation is noted during the inspection. The review can determine the remaining life of the vessel based on the measured thickness and therefore the reviewer can recommend a remedial action such as a repair or alteration and or replacement.

When a pressure equipment is prepared for repair, alteration and /or replacements, a work package including a set of necessary drawings, lists of materials, list of required resources and work schedules should be prepared. This will allow timely sourcing, ordering or fabrication of required parts or section. It must be noted that the repair or alterations must be reviewed and approved by the NBBI Commissioned inspector before the work starts, if the pressure equipment is ASME stamped and registered with the National Board of Boilers and Pressure Vessels Inspectors (NBBI).

Companies operating the boilers must have an efficient system in place for storage and access of inspection reports, drawings, asset review reports to ensure that all data required for reliable assessment are available. The system can be in the form of hard files or preferably electronic files, protected against unwanted and unauthorized changes, revisions or deletions.

Every boiler and pressure equipment must have an inspection plan for future based on their conditions to comply with jurisdictional regulations, company's integrity policy and asset integrity recommendation. Company management is bound to ensure such inspections are carried out and the safety of personnel, property and environment are maintained.

PART ONE

CHAPTER-I   STEAM BOILERS

1.0   INTRODUCTION

A high- pressure boiler is defined by ASME Sec. I, as a steam boiler that operates at a pressure higher than 15 psig or a hot water boiler that operates at a water temperature greater than 250°F or a pressure greater than 160 psig. Low pressure boilers are defined by ASME in Section IV of the BPVC "Rules for Construction of Heating Boilers." It defines low pressure boiler as a steam boiler that operates at a pressure no greater than 15 psig or a hot water boiler that operates at temperatures not greater than 250°F and pressures not exceeding 160 psig.

Steam boilers are used to heat the water to produce steam for use in various applications. Some of the applications are water treatment facilities, power plants, refineries, chemical plants and food industries. There many other applications such as heating system in buildings, fruit processing, pharmaceuticals, metal making factories, dyeing factory, dairies and so on.

1.1 TYPES OF BOILERS

There are three basic types of boiler construction, electric, fire tube and water tube. A fire tube boiler contains the fire and products of combustion inside the tubes and the water and steam is outside the tube. A water tube boiler has the fluids on the other side, tubes surround the water and the fire and flue gas is on the outside of the tubes. Each boiler type may be further classified in accordance to design, output, pressure rating and application

Boilers may be designed for onshore application, locomotive or marine duty either portable (trailer mounted) or stationary. The boilers with limited capacities are supplied as package but for higher capacities they are field-erected.

The power boilers are used for commercial, institutional or industrial applications. Fire tube boilers also may be used for steam turbine generation duty at lower electrical outputs, whereas only WT and nuclear reactor boilers are employed for utility power generation.

For most larger industrial applications around the world, either FT or WT boilers are generally employed.

1.2 ELECTRIC BOILERS

An electric boiler is defined as a power boiler or high temperature water boiler in which the source of heat is electricity. The electrical source can be as high as 15 kV.

Electric boilers with smaller hot water and lower pressure steam generation applications are used in bakery, jacketed kettles, fish pots and other cooking equipment, wineries, breweries, textiles, laundries, humidification, steam baths, clean rooms, and pharmaceuticals and offer significant advantages over fossil-fuel boilers.

Typically, electric boiler designs are limited to under 600 kWh output. However, in some countries where electricity is of particularly low cost, electric boiler designs of up to 12,000 to 24,000 kWh or more are commonly used in large-process industry.

Some of the advantages are:

- Compact designs

- No stack required

- No emission

- Quiet operation

- High Efficiency

- Ease of maintenance

There are two types of electric boilers:

Resistance-Element Electric Boilers

Electrode Steam Boilers

1.3 FIRETUBE BOILERS

In this type of Boilers, the water is fed into the big drum in which tubes are installed horizontally from one end to another and fixed to the tube sheet on either side of the drum. The Burner is installed in one end and the flame is directed inside the tubes. The flue gas may pass one or several passes and heat the water surrounding the tubes.

There are different types and designs of fire tube boilers for various applications however the type that is commonly used in refineries is shown in Fig. 1.1

Firetube boilers come in several configurations and arrangements. Basically, they are cylindrical in shape (Figure 1.1) and are further defined by position and modifications to the general form. The arrangement in is typical of a Horizontal Return Tubular boiler (HRT) which is an early design that has survived to modern times. Return in the label indicates the flue gasses flow down some of the boiler tubes from one end to the other then return through the remaining tubes. Typically the shell of the boiler is extended at the end where the gas makes the turn to form a "turning box" which is closed by large cast iron doors .The doors could be at the front or rear of the boiler depending on how it's constructed relative to the furnace.

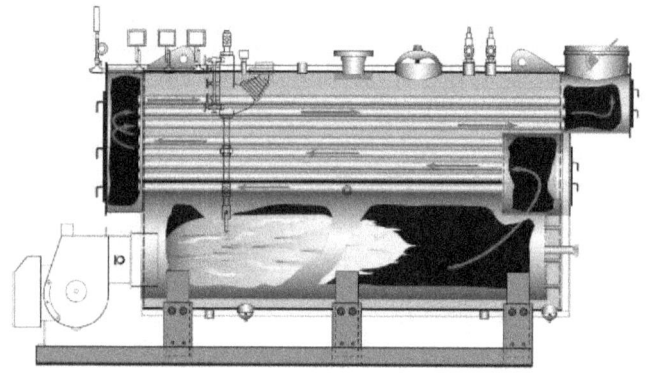

Fig 1.1   Typical Firetube Boiler

The furnace is typically a brick walled enclosure constructed below the boiler. Many Units were built with the brick serving as a base to support the boiler.

Firetube boilers are subdivided into three groups. Horizontal return tubular (HRT) boilers typically have horizontal, self-contained fire tubes with a separate combustion chamber. Scotch, Scotch marine, or shell boilers have the fire tubes and combustion chamber housed within the same shell. Firebox boilers have a water-jacketed firebox and employ, at most, three passes of combustion gases.

Firetube boilers are used for applications ranging from 15 to 3000 hp. The steam pressure in firetube boiler is limited to 450 psig.

Packaged firetube boilers generate low pressure saturated steam, generally below 300 psig.

A locomotive boiler is a firetube boiler (Fig.1.2) modified to provide some water cooling of the furnace. The increased cost of the boiler to create a water jacket around the furnace was justified for locomotive service because the steel and water were considerably lighter than the refractory.

13

Fig. 1.2 Locomotive boiler

1.4 WATERTUBE BOILERS

Watertube boilers convert heat from burning fuel within a furnace chamber to generate either hot water or steam (often at very high pressure and temperature). However, contrary to Firetube boilers, the water is held within the tubes and heat from hot gases flows across the tube walls from the outside.

The benefit of watertube boilers is the fact that, for any given tube diameter, WT boiler tubes have a greater heating surface than FT boiler tubes. This is because the heating surface of WT boiler tubes is the larger outer wall, whereas in FT boilers the heating surface is the smaller inner wall.

Watertube boiler design and construction provide for much greater capacity, pressure, and versatility than FT boilers because of the subdivision of pressure parts and the ability to rearrange boiler components into a wide variety of configurations. The steam output may be from under 1500 lb/hr to several million lb/hr.

Steam drums are used on recirculating boilers that operate at subcritical pressures. The primary purpose of the steam drum is to separate the saturated steam from the steam-water mixture that enters the drum. The steam-free water is re-circulated within the boiler with the incoming feedwater for steam generation. The saturated steam is removed from the drum through a series of outlet nozzles, where the steam flows to a super-heater for further heating.

The saturated steam is pure steam that is at the temperature that corresponds to the boiling temperature at a given pressure.

The steam drum is used for the following purposes:

- To mix the saturated water that remains after steam separation with the feedwater.
- To mix the chemicals that injected into the drum for the purposes of corrosion control and water treatment.

- To purify the steam by removing the contaminants and residual moisture.
- To provide the source for a blow-down system where a portion of the water is rejected as a means of controlling the boiler water chemistry and reducing the solids content.
- To provide storage of water and to accommodate any rapid changes in the boiler load.

The most important function of the steam drum is the separation of steam and water. Separation by natural gravity can be accomplished with a large steam-water surface inside the drum. This is not the economical choice as it results in larger steam drums, and therefore the use of mechanical separation devices is better choice for separation of steam and water.

The watertube boilers are the type used in most industrial plants because of the capacity and safety issues.

WT boilers are safer from explosion than FT boilers because the drum is not exposed to the radiant heat of combustion. If tubes rupture, there is only a relatively small volume of water that can instantly flash to steam.

Inspection, cleaning and maintenance of the boilers is probably comparable for modern WT and FT boilers, but WT boilers tend to require more total hours of operator involvement per year, simply because they are generally physically bigger. Another general rule is that WT boiler systems require more highly trained operators because of their greater complexity.

Watertube boilers are installed in commercial and institutional buildings, smaller industries, large industrial processors, and power generators. In fact, the utility power industry is the single largest user of WT boiler capacity in the world.

Smaller WT boilers are available as shop-assembled packaged units, whereas large boilers are constructed in sections and shipped to site for field erection. The largest field-erected boilers can provide steam, superheated to 1,100 °F (593 °C) and at pressures in excess of 2900 psig.

It is assumed that the readers are familiar with the Boilers and their function therefore the explanation of the operation, functions and the designs are avoided. Only the points which are related to the inspection are discussed.

Fuel oil-fired and natural gas-fired watertube package boilers are subdivided into three classes based on the geometry of the tubes. The (A) design has two small lower drums and a larger upper drum for steam-water separation. In the (D) design, which is the most common, the unit has two drums and a large-volume combustion.

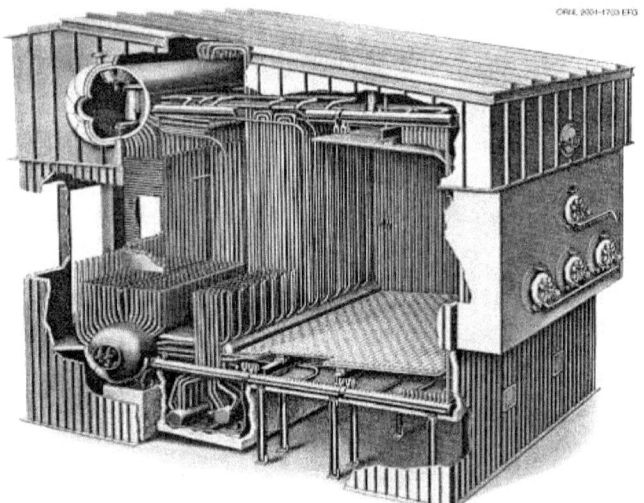

Fig.1.3   typical Water tube boiler

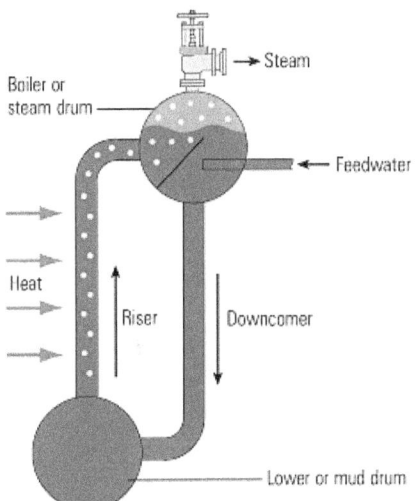

Fig.1.4   sketch of Mud Drum and steam drum

Fig-1.5   Steam Drum with Tube holes

Fig 1.6   Internal view of steam drum

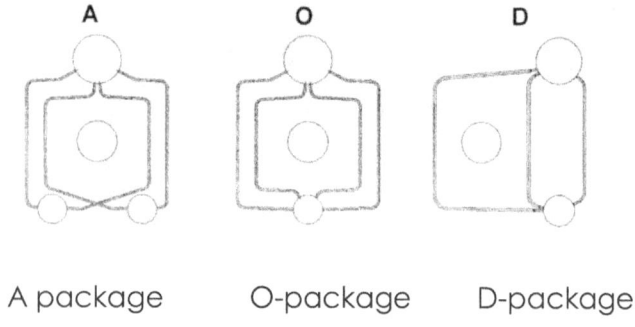

A package   O-package   D-package

Fig.1.7 watertube package types

CHAPTER-II

DEGRADATION OF BOILERS

2.0 COMMON DEGRADATIONS IN BOILERS

The inspectors should be aware of the degradations that may take place in various parts of the Boilers. In other words, Inspector should know what he/she must look for when performing an inspection. There are some of the degradations as a result of corrosion which can be seen when performing Boiler inspection:

2.1 DETERIORATION OF BOILER TUBES

Corrosion of tubes and the drums is largely dependent on the water and water chemistry used within the boiler. Some of the more common types of waterside corrosion include caustic corrosion, hydrogen attack, acid corrosion,

21

oxygen pitting or localized corrosion and stress corrosion cracking.

a)   Deposit inside the tubes

A significant factor in the degree of waterside corrosion is the amount of corrosion product deposited. Deposits restrict the heat transfer and lead to local overheating of the tubes, which can cause concentration of contaminants and corrosives. Depending on which contaminants are present in the feedwater, different deposition locations, rates, and effects will be experienced. See Fig.2.1

Fig. 2.1 Boiler tube fouling
(Courtesy of AONG site)

b) Caustic attack

Localized wall loss on the inside surface of the tubes result in increased stress and strain in the tube wall. Caustic attack occurs when there is excessive deposition on tube inside surfaces. This leads to diminished cooling water flow in contact with the tube, which in turn causes local under-deposit boiling and concentration of boiler water chemicals. Sodium hydroxide can concentrate to high pH levels. At high pH levels, the steel's protective oxide layer is soluble and rapid corrosion can occur. Deposits normally occur where flow is disrupted and in areas of high heat input.

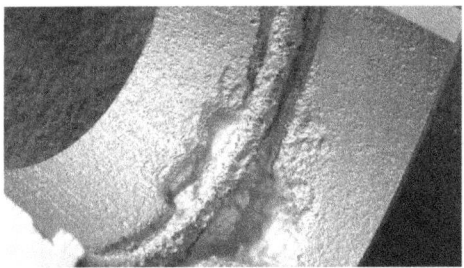

Fig.2.2   Caustic attack on tube

This failure also appears in the form of intergranular attack along the grain boundary causing cracking.

23

(c) Hydrogen Damage

A loss of ductility on tube material or Hydrogen Embrittlement due to hydrogen can lead to a brittle rupture. Hydrogen damage is most commonly associated with excessive deposition inside the tube and a low PH of feed water. When water chemistry is upset, such as what can occur from condenser leaks, particularly with saltwater cooling medium, leads to acidic (low pH) contaminants that can be concentrated in the deposit.

Under-deposit corrosion releases atomic hydrogen which migrates into the tube metal, reacts with carbon in the steel (decarburization) and causes inter-granular separation.

(d) Oxygen Pitting

The aggressive localized corrosion and loss of tube wall near economizer feedwater inlet is common in an operating boiler. Flooded or non-drainable surfaces are most susceptible during outage periods. Oxygen pitting occurs with the presence of excessive oxygen in boiler water. It can occur during operation as a result of ingress of air at pumps or failure in operation of water treatment equipment before feeding the boiler. Pitting corrosion of economizer tubing normally results from inadequate oxygen control of the boiler feedwater.

This also may occur during extended out-of-service periods, such as outages and storage, if proper procedures are not followed in lay-up. Non-drainable locations of boiler circuits, such as super heater loops, sagging horizontal super-heater and re-heater tubes, and supply lines, are especially susceptible.

More generalized oxidation of tubes during idle periods is sometimes referred to as *out-of-service corrosion.* Wetted surfaces are subject to oxidation as the water reacts with the iron to form iron oxide.

When corrosive ash is present, moisture on tube surfaces from condensation or washing water can react with elements in the ash to form acids that lead to a much more aggressive attack on metal surfaces.

For full protection against oxygen pitting during shutdown, the boiler should be kept full of water treated with an oxygen scavenger and blanked or capped with nitrogen.

(e) Acid Attack

Corrosive attack of the internal tube metal surfaces, resulting in an irregular pitted or in extreme cases, a "Swiss cheese" appearance of the tube ID. Acid attack most commonly is associated with poor control of process during boiler chemical cleanings and/or inadequate post-cleaning passivation of residual acid.

(f) Stress Corrosion Cracking (SCC)

The Stress Corrosion Cracking (SCC) is characterized by a thick wall, brittle-type crack. This may be found at locations of higher external stresses like near attachments.

SCC most commonly is associated with austenitic stainless steel super-heater or re-heater materials and can lead to either trans-granular or inter-granular crack propagation in the tube wall. It occurs where a combination of high-tensile stresses and a corrosive fluid are present. See Fig. 2.3

The damage results from cracks that propagate from the inside. The source of corrosive fluid may be carryover into the super-heater from the steam drum or from contamination during boiler acid cleaning if the super-heater is not properly protected.

Fig. 2.3- SCC on tube

g) Corrosion Fatigue

The corrosion fatigue is initiated on internal surface of the tube in the form of a trans-granular cracks which typically occurs adjacent to external attachments.

The tube damage occurs due to the combination of thermal fatigue and corrosion. Corrosion fatigue is influenced by boiler design, water chemistry, boiler water oxygen content and boiler operation. A combination of these effects leads to the breakdown of the protective magnetite on the internal surface of the boiler tube. The loss of this protective scale exposes tube to corrosion.

The locations of attachments and external welds are most susceptible. The problem is most likely to progress during boiler start-up cycles.

2.2 EXTERNAL DAMAGES OF TUBES

Fuel constituents and metal temperatures are important factors in the promotion of fireside corrosion. Fireside corrosion can be classified as either low-temperature attack or high-temperature oil-ash corrosion. Corrosion may occur on the flue-gas side of economizer and air pre-heater tubes. The severity of this corrosion depends on the amount of sulfur oxides or acid in the fuel burned and on the temperature of the flue gas and of the type of fluid being heated. When sulfur oxides are present in the flue gases, corrosion tends to be severe if the gases cool down to the dew-point temperature. The gas temperature in economizers and preheaters should be kept above about 325°F (163°C) to prevent condensation of corrosive liquid. Actual dew point can be calculated from the flue gas composition and should be performed for fuels with high sulfur levels. This may be affected by designing the tubing and the water flow in the tubing so that the gas temperatures are controlled.

The external degradations are listed below:

(a) Fireside Ash Corrosion (Super-heater)

The fireside ash corrosion is a function of the ash characteristics of the fuel and boiler design. It usually is associated with coal firing but also can occur for certain types of oil firing. Ash characteristics are considered in the boiler design when establishing the size, geometry and materials used in the boiler. Combustion gas and metal temperatures in the convection passes are important considerations. Damage occurs when certain coal ash constituents remain in a molten state on the superheater tube surfaces. This molten ash can act as a highly corrosive element.

(b) High-temperature Oxidation

The appearance is similar to fireside ash corrosion. The high-temperature oxidation can occur locally in areas that have the highest outside surface temperature relative to the oxidation limit of the tube material.

To differentiate between these types of degradations, it is better to perform the tube analysis and evaluate both internal and external scale and deposits.

(c) Failure due to sudden overheating

When tube failures occur due to overheating, a careful examination of the failed tube section reveals whether the failure is due to rapid escalation in tube wall temperature or a long-term, gradual buildup of deposit.

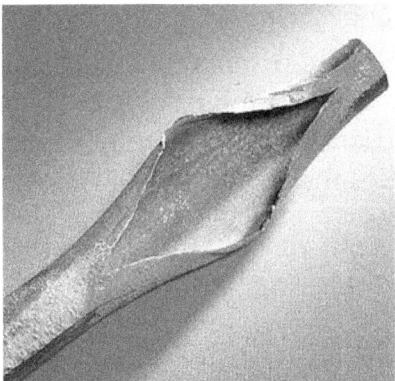

Fig. 2.4   Failure due to sudden overheating

When conditions cause a rapid elevation in metal temperature to 1600°F or above, plastic flow conditions are reached and a rupture occurs. Ruptures characterized by thin, sharp edges are identified as "thin-lipped" bursts (See Fig.2.4)

(d) Water-wall Fireside Corrosion

The external tube's metal wastage leads to thinning and increasing tube strain. The corrosion occurs on external surfaces of tube walls when the combustion process produces a reducing atmosphere.

This is common in the lower furnace of process recovery boilers in the pulp and paper industry. For conventional fossil fuel boilers, corrosion in the burner zone usually is associated with coal firing. Boilers having unadjusted burners or operating with staged air zones to control combustion can be more susceptible to larger local regions possessing a reducing atmosphere, resulting in increased corrosion rates.

(e) Fireside Corrosion Fatigue

Tubes develop a series of cracks that initiate on the outside diameter surface and propagate into the tube wall. Since the damage develops over longer periods, tube surfaces tend to develop appearances described as "elephant hide," "alligator hide" or craze cracking. Most commonly these are seen as a series of circumferential cracks. Usually found on furnace wall tubes of coal-fired once-through boiler designs, but also has occurred on tubes in drum-type boilers.

Damage initiation and propagation result from corrosion in combination with thermal fatigue. Tube OD surfaces experience thermal fatigue stress cycles which can occur from normal shedding of slag, soot-blowing or from cyclic operation of the boiler. Thermal cycling, in addition to subjecting the material to cyclic stress, can initiate cracking of the less elastic external tube scales and expose the tube base material to repeated corrosion.

(f) Short-term Overheat

Failure results in a ductile rupture of the tube metal and is normally characterized by the classic "fish mouth" opening in the tube where the fracture surface is a thin edge. Short-term overheat failures are most common during boiler start up. Failures result when the tube metal temperature is extremely elevated from a lack of cooling steam or water flow. A typical example is when superheater tubes have not cleared of condensation during boiler start-up, obstructing steam flow. Tube metal temperatures reach combustion gas temperatures of 1600°F or greater which lead to tube failure.

(g) Long-term Overheat

The overheating of the tubes in Boilers is caused by creep failure. The failed tube has less swelling with a narrow split along the tube. In the case of short-term overheat the split is wider. Tube metal often has heavy external scale build-up and secondary cracking. Long-

33

term overheat occurs over a period of months or years.

The super-heater and reheat super-heater tubes commonly fail after many years of service due to creep. During normal operation, alloy super-heater tubes will experience increasing temperature and strain over the life of the tube until the creep life is expended.

Furnace water wall tubes also can fail from long-term overheat. In the case of water wall tubes, the tube temperature increases abnormally, most commonly from waterside problems such as deposits, scale or restricted flow.

Overheating is one of the most serious causes of deterioration of boilers. Overheating of the boiler tubes and other pressure parts may result in oxidation, accelerated corrosion, or failure due to stress rupture. Although overheating can occur during normal boiler operations, most often it results from abnormal conditions, including loss of coolant flow or excessive boiler gas temperatures.

These abnormal conditions may be caused by inherently faulty circulation or obstructed circulation resulting from water tubes partly or wholly plugged by sludge or dislodged scale particles. Over-firing or uneven firing of boiler burners may cause flame impingement, short term overheating, and subsequent tube failure. The results may be oxidation of the metal, deformation of the pressure parts and rupture of the parts, allowing steam and water to escape.

(h) High Temperature Creep

High-temperature failure takes place in the Creep temperature range >400 °C. The failure is in the form of blister or larger 'fish-mouth' longitudinal rupture with thick edges. Area around the rupture can have scaly appearance.

The causes for this failure are excessive gas temperature or over-firing during start up, hot gas flowing through an area of low circulation due to plugging or scaling, and heat transfer from an adjacent uncooled component.

(i)   Graphitization

Long-term operation at relatively high metal temperatures can result in damage in carbon steels of higher carbon content, or carbon-molybdenum steel, and result in a unique degradation of the material in a manner referred to as graphitization. These materials, if exposed to excessive temperature will experience dissolution of the iron carbide in the steel and formation of graphite nodules, resulting in a loss of strength and eventual failure. Failure of Carbon Steel tube is brittle with a thick edge fracture.

(j) Dissimilar Metal Weld Failure

Failure is preceded by little or no warning of tube degradation. Material fails at the ferritic side of the weld, along the weld fusion line. A failure tends to be catastrophic if the entire tube fails across the circumference of the tube section.

Dissimilar Metal Weld describes the butt weld where an austenitic (stainless steel) material joins a ferritic alloy material. Failures occurs on the ferritic side of the butt weld. These failures are attributed to several factors: high stresses at the austenitic to ferritic interface due to differences in expansion properties of the two materials, excessive external loading stresses and thermal cycling, and creep of the ferritic material. Therefore, the failures are a function of operating temperatures and unit design.

(k) Erosion

Tube experiences metal loss from the OD of the tube due to erosion. Damage will be oriented on the impact side of the tube. Ultimate failure results from rupture due to increasing strain as tube material erodes away.

Erosion of tube surfaces occurs from impingement on the external surface. The erosion medium can be any abrasive in the combustion gas flow stream, but most commonly is associated with impingement of ash or soot blowing steam. In cases where soot blower steam is the primary cause the erosion may be accompanied by thermal fatigue.

(l) Mechanical Fatigue

Fatigue is the result of cyclic stresses in the component. Mechanical fatigue damage is associated with externally applied stresses.

Stresses may be associated with vibration due to flue gas flow or soot-blowers or they may be associated with boiler cycling stresses. Fatigue failure most often occurs at areas of constraint such as tube penetrations, welds, attachments or supports.

External corrosion of boiler parts may be expected when boilers are out of service for long periods of time. The sulfurous acid formed from the reaction of condensed moisture with the sulfur in ash deposits can cause rapid corrosion of boiler tubes and parts. Also, if a unit remains idle for a considerable length of time, a warm humid atmosphere tends to corrode boiler parts and supports, unless adequate mothballing procedures are followed.

(m) Thermal Shock

Thermal shock is caused by a sudden marked change in temperature either from hot to cold or from cold to hot. The stresses resulting from the sudden unequal expansion or contraction of the different parts may cause distortion only or distortion plus cracking. Thick metals are more susceptible to cracking than are thin ones. The most likely time of temperature shock is during unit start-ups. Heating or cooling rates should be controlled to avoid thermal shock.

Of all the modes of boiler failure thermal shock seems to be the one that can happen at any time. Some boilers can fail due to thermal shock right after commissioning and some boilers may fail after years of operation due to an incident of thermal shock.

It's important to understand exactly how thermal shock destroys a boiler. Thermal shock can destroy a boiler in a single incident or it can take several shocks to produce evident damage. There is a specific combination that must exist for thermal shock damage. First the metal of the boiler (or refractory) must be exposed to a change in temperature that's enough to produce a range of stress in the material.

2.3 MECHANICAL DETERIORATION

Mechanical deterioration of boiler parts can result from a number of causes:

1.   Fatigue from repeated expansion, contraction and corrosion, fatigue from

the combined action of fatigue and corrosion.

2. Abnormal stresses created by rapid changes in temperature and pressure especially in the case of thick-walled drums.
3. Improper use of cleaning tools.
4. Improper use of tube rollers.
5. Settlement of foundations.
6. Excessive external loading from connected piping, wind, earthquake and similar sources.
7. Breakage and wear of mechanical parts.
8. Firebox explosion.
9. Vibration due to improper design or support failure.
10. Improper gaskets that allow steam leak to score the seating surface.
11. Non-weather-tight casing that slows external tube corrosion during extended shutdowns.

The non-pressure parts, including refractory linings of heaters, burners, supporting structures, and casings, may also be damaged from overheating. Usually, such overheating is caused by improper operating conditions or as a result of deterioration of other protective parts. For example, if the refractory lining of a heater is permitted to deteriorate from normal wear, erosion or mechanical damage, it will no longer protect the outer heater casing and structural supports adequately, and such parts may in turn begin to deteriorate rapidly.

2.3.1 Structure Failure

Foundation settlement may be a serious cause of deterioration in boilers because of the severe stress that may be set up in the complicated interconnection of parts, in the external piping, and especially in the refractory linings and baffling. Excessive loads on the boiler by the connection of large pipelines may cause damage to the boiler foundation and pressure parts.

Settlement of foundations may also result from heat transmission from the firebox and subsequent drying of the soil. In zones with seismic activity, earthquakes may cause severe damages. The damage will be somewhat similar to that caused by foundation settlement and may be particularly severe to refractory linings. Vibrations from high and moderate winds, earthquakes, burner operating instability, and high flue-gas flow across tube banks can damage various parts of boilers as follows:

a. Stacks may be so damaged that they overturn.

b. Air and flue-gas ductwork may be damaged resulting in cracks at corners or connections.
c. Expansion joints may crack.
d. Guy lines may loosen or break.

2.4 FAILURE FROM OTHER CAUSES

In addition to the abovementioned causes of failure, other sources also play role in parallel. Boiler's common failures are categorized in four groups:

- Overpressure
- High or low water level
- Structural weakening
- Operator error

To prevent the overpressure, the following instruments are used:
- Steam pressure gages
- Pressure relief valves
- Fusible plugs

Water level is controlled by following instruments and tools:
- Water glasses or gage glasses
- High and low level alarms
- Automatic trips

Gage glasses or water glasses are to be continuously checked by operators.

Weakening of the structure is grouped in three sources:
- Weakening of pressure parts
- Failure of supports
- Mechanical damage

Weakening of pressure parts includes:
- Overheating
- Loss of metal due to corrosion
- Soot blower erosion
- Flame impingement
- Improper combustion

2.4.1 Water treatment

Improper water treatment or the lack of it contributes to most of the failures. The boiler fails because scale builds up on the internal surface of the tube until some metal overheats to rupture. The tube blocks and fails to allow the steam and water to escape. The water then flashes into steam so quickly that it violently blows the boiler apart. A boiler operator should be comfortable with his water treatment

For years we could count on the reports of boiler failures to list low water as the primary reason the boiler failed. Even today, with special systems and all our knowledge, low water always stands out as a significant cause for boiler failures. Taking all the precautions and conducting the regular testing should prevent them but they continue to occur.

It doesn't matter if it's a hot water boiler or a steam boiler, it should have a low water cut-off; steam boilers should have two. In the last century the most consistent reason for a boiler failure, accounting for about one third of the incidents, was loss of water. You should check the cut-offs as often as possible and under different situations to be certain they are reliable. Low water cut offs come in two basic forms, float and conductance. Float operated cut offs, as their name implies, use a float to detect the water level and a lever connected to the float keeps the float in position and actuates the electrical contacts that open to stop burner operation.

The failures of boilers due to low water continues despite the provisions of extra low water cut-offs and regular testing of them. Perhaps one principle reason is the failure to test them regularly, so a problem is detected before a failure occurs.

2.4.2 Deaerator

Boiler feed tanks with heaters and deaerators are another common piece of pre-treating equipment. They have three principal functions, removing oxygen from the boiler feedwater, heating and storing boiler feedwater. In the case of some deaerators the three functions are served by separate tanks, a deaerator and separate storage tank. Both systems remove air from the water but there are variations in equipment construction and differences in how much air is removed.

Neither removes oxygen completely. A boiler feed tank can only remove oxygen to small values. Deaerators, operated properly, will remove oxygen to minimal amounts.

Removal of the oxygen is achieved by raising the temperature of the water. As the water temperature approaches the boiling point the amount of oxygen the water can hold decreases. Heating the water to 180°F reduces the maximum oxygen absorption to less than 2 ppm. Raising the temperature to boiling reduces that to 0.007 ppm. When the water is ready to boil every molecule of water is prepared to change to steam so the water has very little ability to hold dissolved oxygen. The dissolved oxygen forms bubbles of gas in the water. Complete deaeration is not achieved until those bubbles are removed. It's getting the bubbles out that makes the difference in deaerators.

Deaerators are provided in five types, vacuum, flash, spray, scrubber, and tray. A vacuum deaerator is typically a vessel filled with packing and operated under a vacuum. The packing is not like pump or valve packing, it's like fill, loose pieces of ceramic or plastic materials stacked randomly that act sort of like splash blocks so a lot of the water surface is exposed as it tumbles down through the packing. Producing a sufficient vacuum in a vacuum deaerator will bring the water to a saturated condition. For example, pulling a vacuum of 29"Hg (inches of mercury) produces a condition where 79°F water will boil. As long as the water is warmer than the saturation temperature that matches the pressure inside the deaerator it will be at boiling and a little is actually vaporized.

The air and non-condensable gases are removed from the deaerator by the vacuum pump or steam jet ejector, whichever is used. A steam jet ejector will normally discharge to a condenser that uses the remaining energy in the steam to preheat the water before it enters the deaerator. When a vacuum pump is used provisions are made to heat the water and can include any type of heat.

CHAPTER III

INSPECTION OF BOILERS

3.0   PREPARATION FOR INSPECTION

Before the inspection, the tools needed for the work should be checked for availability, proper working condition and accuracy. This includes tools and equipment that are needed for personnel safety. Safety signs should be provided where needed before work is started.

The following tools are needed to inspect any equipment however some equipment may call for a specific tool:

- Portable lights, battery operated
- Flashlight
- Scraper
- Inspector's hammer
- Inside calipers.
- Outside calipers.
- Direct-reading calipers or special shapes.
- Tube caliper or micrometer
- Pocket knife
- Steel rule

- Pit depth gauge
- Angle gauge
- Marking crayon
- Notebook
- Magnifying glass
- Wire brush
- Plumb bob and line
- thickness measurement equipment
- Small mirror
- Magnet
- Digital Camera

When working in a confined space, a second person must accompany inspector to assist him/her with the work, handling of tools and to stay outside of the equipment to call for assistance if anything happens to the inspector. In confined spaces where harmful gas or vapor exists a breathing apparatus should be used by inspector and helper. Both the inspector and helper should have adequate training on working in confined space and the use of the air apparatus.

The following tools should be readily available in case they are needed:

- Surveyor's level
- Plumber's level
- Magnetic-particle inspection equipment
- Liquid-penetrant inspection materials

- Radiographic inspection equipment
- Ultrasonic inspection equipment
- Megger ground tester
- Micrometer (0 in. – 1 in.)
- Electronic strain gauge caliper
- Bore-scope

Inspector should have the copies of all regulatory documents applicable together with client specific procedures to be followed at inspection site. It is highly recommended to review the history of inspections and previous repairs or modifications on the unit. The details of the material, design data and information to be reviewed and recorded in the notebook for reference and reporting. One of the required check points before the inspection is the review and copying of name plate data to ensure all information do match.

It is always a good idea to have a checklist made before the inspection to ensure all sections are covered and no part is missed out.

3.1 SAFETY PRECAUTIONS

Safety precautions must be taken before any equipment is entered. In general, these precautions include but not limited to the isolating the energy sources, check and reduction of confined space temperatures, flow of air, removal of sludge, gases, chemicals and other stream materials before entering.

The presence of toxic residues inside the equipment must be reviewed. Inspectors and maintenance crew who enter the equipment must wear Protective equipment such as coverall, hard hat, gloves, ear protector, face mask, glass and respirator if necessary.

Before the inspector or other crew enters inside the confined space a work permit must be obtained and Lock Out-Tag Out put in place as per plant regulations. A safety officer or operator must enter first to check and confirm hazard free conditions exist and the unit is isolated. When vanadium dust is present, protective apparatus and clothing must be used when internal inspections are performed. This may also be required when safety identifies residues of chemicals that are harmful to human health. Consult all applicable common site-specific, OSHA and other federal, local, and state safety rules and regulations.

It is mandatory to have safety orientation for those inspectors and maintenance crew who are assigned by contractors and are new to the area and may not be aware of the hazards in the plant.

Contractor shall be advised of their responsibilities on safety for all personnel, including their own staffs, property and environment as per company regulation or jurisdictional laws in place.

Plant owners must ensure all personnel including the visitors are trained and briefed adequately on the safe working and they all are familiar with the procedures and instructions in place.

Inspector should have the copies of all regulatory documents applicable together with client specific procedures to be followed at inspection site.   It is highly recommended to review the history of inspections and previous repairs or modifications on the unit. The details of the material, design data and information to be reviewed and recorded in the notebook for reference and reporting.   One of the required check points before the inspection is the review and copying of name plate data to ensure all information do match.

It is always a good idea to have a checklist made before the inspection to ensure all sections are covered and no part is missed out.

3.2 INTERNAL CLEANING OF BOILERS

Steam-drum internals should be inspected before washing to determine any problems, including poor circulation, poor water quality, and low steam purity.

After the initial inspection, the inside of drums and tubes should be washed down thoroughly to remove mud, loose scale or similar deposits before they dry and become hard, more difficult to remove. The washing operation should be carried out from above if possible, to carry the material downward to the blow-off or hand-holes.

A hose with sufficient water pressure or hand tools should be used to remove soft scale and sludge. The blow-off line should be disconnected prior to the washing process to keep mud and scale out of the blow-down drum. The tubes of horizontal-return-tube boilers should be washed from below and above. It is especially important to ensure that all tubes and headers are clear of sludge after the wash is completed. Water should be passed down each individual tube and observed to exit from below.

Each header, if any, should be opened sufficiently to view the inside and ensure that all sludge has been removed. Precautions should be taken so that the water does not come into the contact with the brickwork of the combustion chamber. If contact cannot be avoided, the brickwork should be dried out carefully when the boiler is fired up.

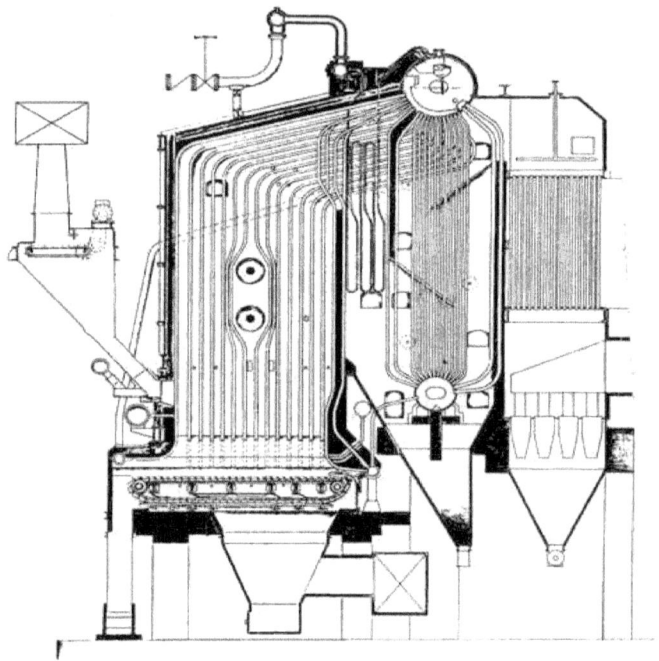

Fig. 3.1 Typical Water-Tube Boiler

The use of an inhibited acid solution on the inside of the boiler is a common method of cleaning the interior surfaces. Prior to cleaning, samples of sludge and deposits should be analyzed to ensure the cleaning solution can adequately remove the material. During the cleaning operation, corrosion probes/coupons are often used to monitor the corrosivity of the circulating solution. After acid cleaning, the interior of the boiler must be neutralized, washed down, and refilled with water. If a nitrogen purge is used after acid cleaning for moth balling, the drums should be checked for oxygen content before entering for inspection.

Acid cleaning should not be used on super-heater tubes, which contains pockets that cannot be thoroughly flushed out. Precautions must be taken to make sure that all sludge is removed after an acid wash.

It is a normal practice to fill pendent-type super-heaters with condensate or demineralized water and to keep the super-heater full of this water while the remainder of the boiler is acid cleaned. During chemical cleaning, all phases of the operation should be closely supervised by experienced, responsible individuals.

All electric power and other ignition sources near boiler must be turned off, during chemical cleaning, to prevent explosion of the hydrogen and other hazardous gases that are normally given off during the cleaning.

3.3 INSPECTION OF BOILERS

3.3.1 Initial data review

Boiler inspectors should be familiar with the types of boilers used in oil and gas industry. Each type of boiler has specific design and operation. The inspection methods will almost be the same however the findings depend on the type of boiler.

The inspector should also be familiar with the process of the boilers in order to understand the failures modes explained above or design related problems. The inspectors are encouraged to talk to the operators in the control room and learn about their problems and the history of the failures, repairs or modification.

Some of the questions to an operator would be:

- Do you experience large fluctuations in feedwater flow during "steady-state" operation?

- Does feedwater flow start and stop frequently during start up?
- During overnight shutdowns, is there frequent "topping-off" of the drum to maintain the level?

Another type of question to the operator would be if the level indictors on steam drum and set levels in control room match. Often the high-pressure (HP) drums with water-level marks that are several inches high, this means there is disagreement between the indicated and actual drum levels. The issue here is that if the operators experience a high-drum-level event, they may think they're operating below the "high-high" level (HHL) but may be above it and drum water may be carrying-over into the HP super-heater.

Also ask if the rates of pressure increase and/or decrease during start up and shutdown exceed those mentioned in the Boiler Manual." If "yes," you should be concerned about the possibility of fatigue damage at manways and thick nozzles and carefully check those locations when the unit is out of service.

When inspecting a Waste Heat Boiler, the first things to investigate is spray-valve behavior. Is spray-valve position near constant at steady load or does it fluctuate—sometimes closing entirely? If your spray valve is continually hunting—that is, opening and closing dozens of times each hour—the hot/cold cycles eventually will stress de-super-heater nozzles to fail and possibly create other damages as well. Control-logic adjustments generally can rectify this condition. This type of information on operations can be available to inspector only when he talks to the operator.

If the de-superheating stations are properly designed and equipped with thermocouples (TCs) upstream and downstream, verify that both are reporting about the same steam temperature when the control system indicates that the spray valve is closed. If the downstream TC is reading lower than the upstream one, water probably is leaking by the "closed" valve. Inspector should give this immediate attention. Concern is that during start ups and shutdowns, water can dribble into the steam piping, collect along the bottom of the pipe, migrate downstream to Super-heater and/or re-heater upper headers, and then run down a few tubes. Cooling effect of the water causes those tubes to contract. Resulting stress can cause tube bowing and tube-to-header weld cracks.

Also check plant data to identify times when spraying to less than the saturation temperature of steam in the pipe plus 30 deg. F. Do a thorough review: start up, steady-state full load with and without duct firing, steady-state low load, and during transients. De-superheating to near-saturation temperature creates the potential for damage like that described above for a leaking spray-water valve.

In watertube boiler, inlet feed water flows through the tubes and enters the boiler mud drum. The circulated water is heated by the combustion gases and converted into steam at the vapor space in the steam drum. These boilers are selected when the steam demand as well as steam pressure requirements are high as in the case where steam is needed to run a steam Turbine.

Fig. 3.1 Picture of a boiler unit

In firetube boiler, hot gases pass through the tubes and boiler feed water in the shell side is converted into steam. Firetube boilers are generally used for relatively small steam capacities and low to medium steam pressures. As a guideline, firetube boilers are competitive for steam rates up to 12,000 kg/hour and pressures up to 18 kg/cm2. Fire tube boilers are available for operation with oil, gas or solid fuels. For economic reasons, most firetube boilers are nowadays "packaged" and manufactured and assembled in shop for all fuels.

The packaged boiler is so called because it comes as a complete package. Once delivered to site, it requires only the steam, water pipe work, fuel supply and electrical connections to be made for it to become operational. Package boilers are generally of shell type with firetube design so as to achieve high heat transfer rates by both radiation and convection.

The firetube boilers are classified based on the number of passes - the number of times the hot combustion gases pass through the boiler. The combustion chamber is taken, as the first pass after which there may be one, two or three sets of firetubes. The most common boiler of this class is a three-pass unit with two sets of firetubes and with the exhaust gases exiting through the rear of the boiler.

Wherever the waste heat is available at medium or high temperatures, a Waste Heat Boiler can be installed economically. Wherever the steam demand is more than the steam generated during waste heat, auxiliary fuel burners are also used. If there is no direct use of steam, the steam may be let down in a steam turbine-generator set and power produced from it. It is widely used in the heat recovery from exhaust gases from gas turbines and diesel engines.

3.3.2 Inspection Requirements

The requirements governing inspection of boilers can differ widely from one location to another since they are often regulated by jurisdictions. Under some jurisdictions, inspections must be made by state, municipal, or insurance companies. Other jurisdictions may allow inspections by qualified owner/user inspectors. In either case, the inspector is usually commissioned by the regulatory authority and must submit reports of the inspection to the official responsible for enforcement of the boiler law. If the boiler is insured, inspection by the insurance company inspector also serves to satisfy his/her company that the boiler is in an insurable condition.

In the absence of a regulating body, the owner- user inspectors or certified external inspectors must be assigned to perform the inspection and prepare the report.

Normally, governmental and insurance company inspectors concern themselves only with the pressure parts of the boiler, the safety valves, level indicators, pressure gauges and feedwater and steam piping between the boiler and the main stop valves, super-heaters and economizers.

The plant inspector should also be concerned with related non-pressure parts, including the furnace, burners, flue-gas ducts, stacks, and steam-drum internals since these can affect equipment reliability and performance.

When inspection by an outside agency is necessary, joint inspections by the outside inspector and the plant inspector can reduce the length of boiler outages and result in shared learning. The outside inspector is primarily interested in ensuring the minimum legal safety requirements are met. The plant inspector should be interested not only in safety but also in conditions that affects reliability and efficiency.

The outside inspector has an opportunity to examine many boilers that operate under widely varying conditions and often can offer valuable advice on the safe operation of the boilers.

3.3.3 Inspection of tubes

Tubes in watertube boilers should be inspected externally from inside the combustion chamber (Firebox) for the following conditions:

a. Sagging or bowing
b. Bulging
c. Oxidation or scaling
d. Cracking or splitting
e. External corrosion
f. External deposit
g. External pitting.
h. Leaking rolls.
i. Overheating (hot spot)

Inspector should gather required data and information on the past operation for the boilers with the history of all modification and repairs. He/she should also review the drawings, Process Flow Diagram and previous inspection reports.

With all required information in hand only, inspector can start the inspection.

Because of the arrangement of the tubes and refractory walls in watertube boilers, visual inspection of the external surfaces of the tubes is usually restricted to the fireside of the tube-walls. Special attention should be given to the following locations:

In watertube boilers, the area from the firebox floor to approximately 10 ft (3 m) above the firebox floor where burners flame exist should be carefully and visually inspected to detect the sign of flame impingement to the tubes and refractory lining.

Use flashlight to detect any sign of leakage or deposits on the tubes. Access is required to all levels for close examination.

The refractory layers at all level to be inspected for cracks, opening or falling sections. In high capacity steam boilers, which are used is petrochemical complexes and power industries, suitable scaffolding must be arranged before the start of the inspection to allow access for inspection of higher levels.

Mud drum should be cleaned from any debris and water for entry. The tube holes, rolled or welded tubes to drum should be carefully inspected for any sign of corrosion, metal loss or cracks. Any corroded section to be photographed and measured with a ruler as a scale. Reading should be recorded.

The tube holes on drum shell to be numbered carefully for traceability of any defective tube to avoid the confusion and the work on wrong tube. The steam drum(s) should also be inspected, and all observations recorded.

If any leaking tube detected or suspected, the tube can be individually tested by water and suitable plugging of both ends in mud and steam drums with water injection connection, vent and pressure gauge.

Most of the times, it is not possible to remove the defective tube or replace them as this require sacrifice of many tubes and removal of refractory layer and sometimes removal of boiler external walls to access the work area. The leaking tube can be plugged at both ends using similar material to avoid galvanic corrosion.

If the numbers of leaking or defective tubes are not much to affect the efficiency of the boiler, the tubes are normally plugged at both ends and left in place.

The Safety Relief Valve(s) installed on the boiler or on the header should be taken to workshop for inspection, testing and calibration. Inspector should witness the operation and ensure that the seating faces are free from any corrosion, pitting prior to calibration. In case of shallow pitting, the seating surfaces may be machined and lapped for full contact.

3.3.4 Inspection of Drums

Thickness measurement should be carried on Mud Drum and Steam Drum at the same points that previously measured to provide a means for calculation of corrosion rate. Tubes should also be subjected to thickness gauging at available accessible locations to detect the corroded areas.

It is always beneficial to inspect the internal surface of the tubes at either side with borescope to detect the amount of deposits inside the tubes. This information can provide required data to decide if chemical cleaning is required for the removal of the deposit. The amount of the deposit also provides information on the quality of feed water and chemical treatment.

Steam drum provides the access to the water/steam side of a boiler. That's an important point. All boiler water eventually passes through one or more steam drums in a triple-pressure unit, so a great deal can be learned from a proper inspection.

Inspector's checklist should include the following:

- Inspect the drum internal surface.
- Investigate any signs of distress on steam separation equipment and the crack which are caused by thermal stress or corrosion. If components are worn, measure metal thickness to quantify the extent of material loss.
- Verify mechanical integrity.
- Determine if the actual waterline is at the level designed.
- Scoop-up and carefully examine drum debris. Run tests to characterize the debris if necessary.
- Check the manway ring to see if machining is necessary to restore it to the desired flat condition.

A thorough examination of the drum internal surface demands that inspectors get up close to the metal and look for pits, nicks, cracks, tubercles, corrosion, erosion, etc. Take notes and pictures of anything that looks out of the ordinary. Assess overall drum condition, tracking surface defects from manufacture, such as those created while rolling steel for the HP drum.

Has smoothness/roughness of the drum surface changed during the inspection interval?

For example, perhaps tubercles have formed. It is particularly important to identify any pitting attack and to inspect what the pits look like underneath. Keep in mind that pits may corrode under pits, resulting in pits larger than those at the surface.

Look at the steam separators and the belly pan for the following:
- Cracks in the belly pan (or baffle plate)
- Cracks in the corners of the final separator
- Thinning caused on the belly pan, as evidenced by the loss of material or the hole in the LP belly pan.
- Early corrosion attack on the LP cyclone and advanced attack (huge hole) on the cyclone.

Ultrasonic (UT) thickness measurement is useful for recording the condition of drum internals, tubes and piping. This can be used for calculating the rate-of-wear and timely maintenance planning.

Thickness testing of internals compares original specifications of drum components to data collected during the inspection. More sophisticated ultrasonic testing such as shear-wave and phased-array (PAUT)can be used to identify cracking, if suspected.

Look for the obvious defects such as broken "U" bolts, missing nuts and other hardware, plugged chemical-feed lines and separators inside the drum.

Sometimes the water level falsely appears lower than actual because the level transmitter is located too near to a down-comer or pump intake. Their suction effect "pulls" on the level sensing lines. Plugging of sensing lines is another possible cause of false readings or sluggish response. The primary cause of faulty level indication is due to an instrumentation that is out of calibration

3.3.5 Non- Destructive Testing

The intrusive inspection techniques should be considered to more accurately assess boiler condition. This might include one or more of the following:

- Ultrasonic testing (UT) of tubes and piping beyond spot checks; select optimum technology from among A-scan, phased-array or shear-wave as specific tasks dictate.

- Dye-penetrant (DPT) or magnetic-particle (MPI) testing of suspect areas.
- Removal and inspection of de-superheaters.
- Borescope examination.
- Sampling and analysis of pressure-part materials

An ultrasound transducer in contact with a metal by a couplant, sends the wave into the material. Reflection off the back wall or imperfection records the wall thickness or depth of the discontinuity. A-scan UT typically is selected for thickness testing, especially where erosion or external corrosion has been identified. It's easy to use, is fast and accurate.

Wall thinning generally is experienced first after tube bends near the upper headers. To access the tube-bend area, remove header baffles as necessary. Check pipe elbows (jumpers) at the apex of the elbow; and straight runs downstream of control valves and orifice plates, especially where temperatures are more than 280F and less than 340F.

Phased Array Ultrasonic testing (PAUT) makes use of multiple fixed-angle transducers in a single probe to quickly and accurately characterize flaws to provide the depth of indications, which show up as echo blips on the instrument's screen. It has become the instrument of choice in inspections for its ability to identify even small flaws in complicated geometries.

Phased-array instruments and probes are more complex and expensive than conventional UT and technicians require more experience and training to use it. However, compared to radiography (RT), phased-array ultrasonic testing offers several advantages, including:

- No work stoppage or welder relocation is required to check the work because there are no safety hazards with UT as there are with RT (radiation).
- Inspections are conducted faster.
- Better suited for the detection of planar, crack-like indications, such as lack of fusion which are conducive to premature failure.
- The exact depth of a defect is revealed, facilitating its removal and rework.
- Digital RT is making great strides to eliminate it.

Shear-wave UT is about two to three times less expensive but does not give depth of indication. Best applications for it are straight pipe, elbows and circumferential welds.

Dye Penetrant (DPT) and Magnetic Particle (MPI) methods are used to identify *cracks at the surface* of the tube material at or near a weld. They also are used for regular checking of tube/header joints at the economizer water inlet when more than 500 start/stop cycles have been accumulated; to verify the integrity of super-heater and re-heater tube/header joints where warped tubes have gotten worse over time; and to pinpoint suspect leaks in vent and drain lines.

Dye Penetrant testing (DPT) is a low-cost and widely applied inspection method for locating surface-breaking defects in all non-porous materials (metals, plastics, and ceramics). For inspection of ferrous components, however, MPI often is preferred because of its subsurface detection capability.

Dye-penetrant testing is used to detect casting and forging defects, cracks, and leaks in new products, as well as cracks on in-service components not visible to the naked eye. The red dye is used for a high contrast against a white developer background. The developer draws out the penetrant from the flaw over a wider area than the real flaw, enhancing its visibility.

The most common MPI method relies on finely divided iron or magnetic iron oxide particles held in suspension in a suitable liquid (often kerosene). This fluid is referred to as the "carrier." The particles often are colored and usually coated with fluorescent dyes that are made visible with a hand-held ultraviolet (UV) light.

The particle suspension is sprayed or painted over the magnetized specimen to localize areas where the magnetic field has protruded from the surface. The magnetised particles are attracted by the surface field in the area of the defect and hold on to the edges of the defect and define it by a build-up of particles.

MPI is used to inspect machined parts before they are placed in service. It is also used to inspect the parts in service for fatigue cracking. The testing method is easy to apply and takes less time than UT and DPT, but it does not work well with complex geometries.

Borescope inspection was used for inspection of gas turbines, steam turbines and generators. Its usage on other parts of boilers referred below has been adopted now:

- In Economizer, cold end, when pitting or cracking of the tube internal surface is of concern because of cyclic stresses and/or water-quality issues.

- In Economizer, hot end, if damage or deposits are suspected from water-quality excursions in conjunction with excessive steaming.
- In HP evaporator, hot end, if water-chemistry excursions make internal deposits a risk or iron levels are above 1 ppm.

- In LP evaporator tube rows are most susceptible to corrosion damage based on a circulation analysis or failure history.
- In First tube bundle of the superheater, downstream of the steam drum, if deposits from carryover are suspected because of a steam-separator failure.

Tube samples are taken to characterize both hard scales that form in LP boilers and magnetite and copper deposits in HP boilers. Such deposits reduce heat transfer and efficiency and cause the Heat Recovery Steam Generator (HRSG) more susceptible to overheating and tube failures.

The sampling of the tubes must be avoided as much as possible if NDT method can provide the required data. Tube sampling is difficult, expensive and has obvious inherent risks.

Laboratory analysis of the deposits may suggest corrections to water chemistry and operating procedures and the most efficient method for removing deposits. Base metal analysis will advise on the condition of boiler tubes and whether expectations of service life must be revised downward.

3.3.6 Wall Thickness Measurement

The wall thickness of the tubes in a boiler is to be monitored. Thinning mechanisms can be identified and monitored through wall thickness measurements.

The methods used to determine the wall thickness of tubes is ultrasonic testing method and radiography. If the tube is accessible to measure the inside and outside diameter, then the thickness is that point can be calculated. There is also a destructive method where a piece of tube is cut and inspected visually for corrosion sign and for metallographic test.

Some Thickness monitoring locations (TMLs) can be selected for measurement of the thickness during the outage or shutdown to provide a means to determine the deterioration extent and to calculate the deterioration rate. This usually involves, placing TMLs marking on all tube passes throughout the firebox. Attention should be made to tubes in economizer and super-heater coils.

Thickness measurements should be recorded and compared to historical readings in the same locations. These wall thicknesses provide a record of the amount of thickness lost, the rate of loss, the remaining corrosion allowance, the adequacy of the remaining thickness for the operating conditions, and the expected rate of loss during the next operating period.

The ultrasonic method for obtaining tube-wall thickness is the common method. Proper cleaning of the external surface oxidation or compensating for the thickness of the oxide layer is essential to properly assessment of metal loss rates.

If a thin area is detected, follow-up inspection using spot ultrasonic or radiographic inspection is done.

Another method is the use of guided wave testing where an acoustic wave is introduced into the tube that travels either axially or longitudinally along the tube to detect the defective areas, then follow-up inspection using spot ultrasonic or radiographic methods is done to confirm whether or not flaws truly exist at the identified locations.

When the boiler is under inspection, it is the best time to perform the inspection of water Deaerator, if it is out of service.

Normally steam is used to deaerate the feed water from oxygen, therefore the escaping steam from deaerator contains oxygen which most likely causes the corrosion of the shell plate around the vent nozzle. Thorough inspection of this area and the drain nozzle of the deaerator vessels are of prime importance and highly recommended.

The non-pressure items such as boiler steel structure, pipe rack, foundation, earthing connection, insulation, stacks, supports, clamps and control system shall also be inspected. If any part required detail inspection and review, inspector may recommend a survey by discipline engineer.

The inspector shall prepare a detailed report on completion of the inspection and prior to re-start of the boiler to provide adequate time to engineers for the review of the detected defects and decide on the type of any repair or replacement to ensure the integrity, safety and reliability of the system

Maintenance outages provide an opportunity to gain access to the tubes and other internals to assess their present condition and allow for data to be obtained and to predict their future reliability.
Inspections that can be performed during outages include:

a. Visual examination.
b. Wall thickness measurements.
c. Tube diameter or tube circumference measurements.
d. Tube sagging or bowing measurements.
e. Pit depth gauging.
f. Radiography (digital radiography).
g. Hardness measurements.
h. Borescope/video probe.
i. In-situ metallography/replication.
j. Dye penetrant testing.
k. Magnetic particle testing.
l. Tube section removal for creep testing.
m. Tube section removal for metallography.
n. Tube removal for detailed visual examination.
o. Testing of tube skin thermocouples.
p. Inspection of Refractory

3.4 Inspection of Heat Recovery Steam Generator

In addition to general physical inspections mentioned for common boilers in above paragraphs, some important points are mentioned here to encourage the inspector to pay more attention to process side.

Inspector is strongly suggested to compare the water temperature at the economizer outlet to the saturation temperature of water in the steam drum at the exact same time. If the two are within a couple of degrees, that indicates the water exiting the economizer contains steam. To solve this operational problem, the operators must ensure that the start up is as short as possible and maintain economizer feed-water flow. Venting appropriately during filling and start up also is important and the operator should note that.

The inspector is to keep in mind that air pockets in the system block the water flow and once a circuit is blocked it will steam up, making it even more difficult to "clear" during start up. If the temperature of the water at the economizer outlet makes several sudden jumps upward during start up from a lower-than-expected value, this generally means a vapor pockets are being cleared and normal operation is being restored. However, such clearing events, similar to water hammer, stresses the economizer unnecessarily.

Review data to identify any rapid temperature change in the water inlet of Economizer (30 F/min, plus or minus) when the unit is in operation or offline. A rapid change in the temperature is viewed as a thermal shock to the economizer. Also verify that the inlet temperature doesn't drop below 120F. Low temperature can cause the dew point condensation and tube wastage.

Verify with operators that the Economizer recirculation pump is operating early in start up and *before* new feedwater is added to the LP economizer. This allows the recirculating flow to buffer the temperature difference between the hot tube panels and the relatively cold water being injected into the circuit.

When the recirculating pump starts, listen for audible signs of cavitation. The repeated cavitation can cause pump damage, that is the reason some operators do not use the recirculating pumps to avoid the cavitation and this can damage the economizer by thermal shock/or dew-point corrosion.

PART II

CHAPTER IV

HEAT EXCHANGERS

INTRODUCTION

A heat exchanger is a device used to transfer the heat between two or more fluids at different temperature and thermal conditions. Heat exchangers are devices used to transfer energy between two fluids at different temperatures. They improve energy efficiency, because the energy already within the system can be transferred to another part of the process, instead of just being pumped out and wasted.

4.0 TYPES OF HEAT EXCHANGERS

4.1 PLATE HEAT EXCHAMGER

Plate heat exchangers have become the most popular among the rest due to their small size, easy cleaning, quick assembly and with minimal hydraulic resistance.

The heat exchanger works according to the cross scheme. Sections in turn are filled with heated and cooled medium. Through the plates, heat exchange takes place. Seals of various shapes provide filling sections.

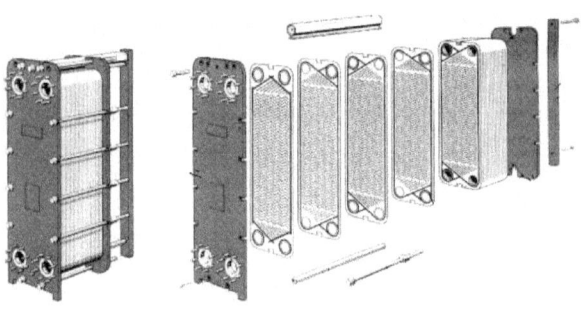

4.1 Plate heat exchanger

4.2   SHELL AND TUBE HEAT EXCHANGER

The most common heat exchangers are Shell and Tube type exchangers. It consists of a shell, a tube bundle, a channel head, floating head cover and shell cover.

Various types of S/T Heat Exchangers are shown below:

4.2.1   Straight-Tube Design

This design is used in heavy fouling fluids service. The head assemblies can be removed to facilitate the cleaning of tubes mechanically. The ability to handle large temperature differences between the fluids may be limited and depends upon the tube sheet configuration i.e. fixed or floating.

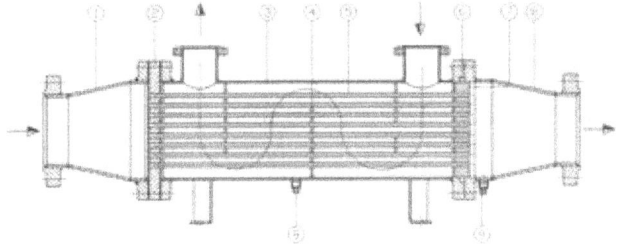

Fig. 4.2 Typical Shell and Tube Heat
          Exchanger

4.2.2 U-Tube Design

The U-tube design consists of straight length tubes bent into a U-shape.
The bundle is fitted with tube supports or flow baffles, depending on the fluid flowing outside the tubes. The tube assembly is placed in a shell to contain the fluid on the outside of the tube bundle.

A head assembly is bolted to the shell to direct the fluid into the tube bundle. The head assembly contains one or more partitions for controlling fluid flow velocity and therefore, the heat transfer coefficient and pressure drop. The U-tube construction allows for large temperature differences between the tube-side and shell-side fluids with the U-tubes expanding or contracting independently of the shell assembly.

U-tube heat exchanger

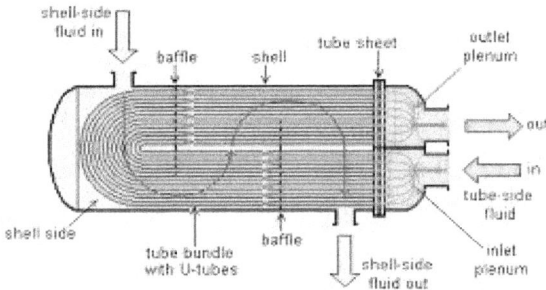

Fig. 4.3   U- Tube Shell and Tube Heat Exchanger

The disadvantage of the U-tube exchanger is that tubes are difficult to clean internally and replacement of leaky tubes in the inner rows involves unnecessary cutting of good tubes in the outer rows. The U-tube bundle exchangers usually have a welded shell cover.

4.2.3   Floating Head Exchangers

This type of exchanger consists of a cylindrical shell flanged at both ends, a tube bundle with a tube sheet at each end, a channel with cover, a floating head cover and a shell cover. The diameter of floating tube sheet is smaller than the shell diameter so that tube bundle can be inserted into the shell. The channel is bolted onto the shell flange to hold the stationary tube sheet in position. Similarly, the floating head cover is bolted onto the floating tube sheet. The shell cover is thereafter bolted in its place.

Suitable partition arrangements in the channel and the floating head cover can provide several tube side passes. The desired fluid flow pattern through the shell is directed by baffles. The floating tube sheet can move in the shell. This type of construction permits free expansion and contraction with changes in temperature.

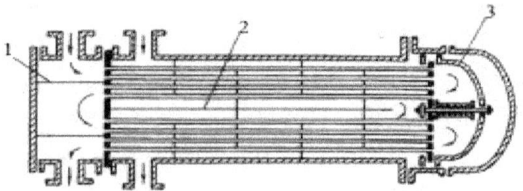

Structure of floating-head heat exchanger

1-baffle in tube pass; 2-baffle in shell pass; 3-floating head

Fig. 4.4 Floating Head Shell and Tube Heat Exchanger

4.2.4 Fixed Tubesheet Exchanger

It consists of two tubesheets welded to the shell with the tubes rolled into the tubesheets, with channel head on either side. Since the tube bundle cannot be pulled out, this type of exchanger is suitable for clean fluid services where there is little possibility of fouling on the outside of the tubes, otherwise chemical cleaning will need to be done. Also, temperature conditions should be such that the stresses due to differential thermal expansion between shell and the tube do not over-stress shell or tube, otherwise expansion bellows will need to be provided on shell to take care of this differential thermal expansion.

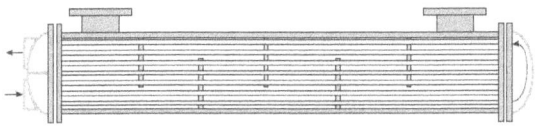

Fig. 4.5 Fixed Tubesheet Heat Exchanger

4.2.5 Re-boiler / Kettle Type Heat Exchanger

The primary use of this exchanger is boiling the fluid for distillation. The kettle shell is used in the re-boilers, or chillers. The fluid to be heated is in the shell and the heating medium, generally steam or process fluid, is in the tubes. Such re-boilers are called process re-boiler. The shell of the kettle type re-boiler has large vapor space over the tube bundle.

A kettle type re-boiler has several advantages over standard heat exchangers in similar service. It has a lower pressure drop and can handle fluctuating load. The same type of construction is used in some chillers. For this service, a volatile cooling medium such as propane is in shell and fluid to be cooled in tubes. The latent heat of vaporization is absorbed from the cooled medium. The bundle may be "U" tube type or straight tube with floating head.

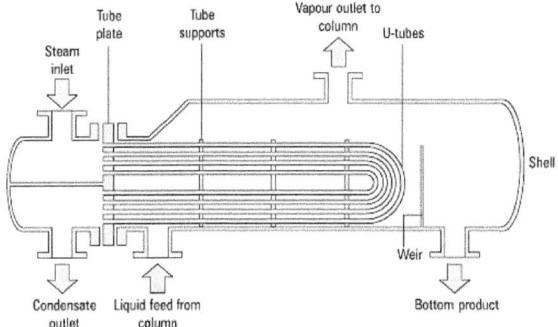

Fig. 4.6   Kettle Type Reboiler

CHAPTER-V

DETERIORATION IN HEAT EXCHANGERS

Deterioration may be expected on all parts of the heat exchangers in contact with hydrocarbons, chemicals, sea water, fresh water, steam and condensate. The form of the attack may be electrochemical, mechanical or combination of both. The degradation may further be accelerated by certain elements such as temperature, stress, fatigue, high velocity of flow and impingement. Various forms of the degradations are explained below:

5.1 Degradation of Shell and Shell Cover

Carbon steel shells are prone to internal corrosion and pitting when hydrocarbon streams contain compounds of Sulphur such as hydrogen sulphide or mercaptan.

At high temperatures hydrogen sulphide reacts with carbon steel and forms Iron Sulphide scales. This usually results in a uniform loss of metal. This type of corrosion is more predominant in preheat exchangers.

The Internal corrosion can also occur due to low temperature hydrochloric acid or hydrogen sulphide corrosion in presence of moisture. Overhead condensers in crude distillation, vacuum, visbreaker and FCC units are prone to this type of attacks.

A combination of wet hydrogen sulphide and hydrochloric acid (that form due to hydrolysis of chlorides in crude during distillation) aggravates the internal corrosion of overhead condenser shells. It will be most pronounced in the bottom part of the shell and lower nozzles. This type of corrosion is uniform or in the forms of a groove following the line of flow of the condensate.

Reboiler shells are prone to internal pitting or grooving due to steam condensate corrosion. Overhead Condensers, Coolers and Exchangers in sour gas and Mono- Ethyl Amine(MEA) / Di-Ethyl Amine(DEA) service are prone to shell side cracking due to stress corrosion cracking phenomenon at the weld joints if they are not properly stress relieved.

Erosion / corrosion will take place around outlet nozzles of cooler shells due to solid particles, if present in stream.

Grooving and thinning of shell may take place in coolers or condensers at the baffle resting locations due to galvanic corrosion.

Pitting type corrosion will take place in carbon steel heat exchangers shell in high temperature MEA / DEA or phenol service.

External corrosion of shells may result due to water seepage in the thermal insulation having high chloride concentration in areas such as offshore installations or coastal facilities.

Carbon steel shells are prone to internal corrosion and pitting when hydrocarbon streams contain compounds of Sulphur such as hydrogen sulphide or mercaptan.

At high temperatures hydrogen sulphide reacts with carbon steel and forms Iron Sulphide scales. This usually results in a uniform loss of metal. This type of corrosion is more predominant in preheat exchangers.

The Internal corrosion can also occur due to low temperature hydrochloric acid or hydrogen sulphide corrosion in presence of moisture. Overhead condensers in crude distillation, vacuum, visbreaker and FCC units are prone to this type of attacks.

A combination of wet hydrogen sulphide and hydrochloric acid (that form due to hydrolysis of chlorides in crude during distillation) aggravates the internal corrosion of overhead condenser shells. It will be most pronounced in the bottom part of the shell and lower nozzles. This type of corrosion is uniform or in the forms of a groove following the line of flow of the condensate.

Reboiler shells are prone to internal pitting or grooving due to steam condensate corrosion. Overhead Condensers, Coolers and Exchangers in sour gas and Mono- Ethyl Amine(MEA) / Di-Ethyl Amine(DEA) service are prone to shell side cracking due to stress corrosion cracking phenomenon at the weld joints if they are not properly stress relieved.

Erosion / corrosion will take place around outlet nozzles of cooler shells due to solid particles, if present in stream.

Grooving and thinning of shell may take place in coolers or condensers at the baffle resting locations due to galvanic corrosion.

Pitting type corrosion will take place in carbon steel heat exchangers shell in high temperature MEA / DEA or phenol service.

External corrosion of shells may result due to water seepage in the thermal insulation having high chloride concentration in areas such as offshore installations or coastal facilities.

5.2   Degradation at Tubes

Turbulent flow in cooling water can cause damage to the protective film on the tubes, which will accelerate electrochemical corrosion between the regions where the film has been removed, and the regions still protected. When the cooling water flows at high velocities, corrosion holes appear in the tube walls along flow lines, promoting holing. Horseshoe-shaped holes are formed open in the direction of flow, with corrosion products stripped from the corroded holes in a typical erosion pattern.

Turbulent flow occurs commonly near the cooling water inlet. This can cause erosion–corrosion in that region. When this erosion–corrosion occurs near the cooling water inlet, it is often called inlet attack.

Deposit attack is a type of corrosion caused by foreign materials entering the tube, and can be broadly grouped into three types:

a) Attack by fixed deposit (local corrosion around retained foreign material). This is localized erosion caused by a build-up of material such as a shell or piece of wood lodged in the tube, eventually causing turbulent flow. Organisms such as barnacles and mussels can also cause deposit attack.

b) Crevice attack under deposits When mud or sand deposits build up inside the tube, corrosion may be caused by battery effects between the region under the deposits and the surrounding region.

c) Vibrating deposit attack When flexible materials such as straw, vinyl or seaweed vibrate in the water flow and repeatedly rub against the tube wall, the protective film may be damaged, leading to corrosion.

Corrosion from polluted seawater- Alloys which offer excellent corrosion resistance in clean seawater can exhibit completely different dynamics in polluted seawater. Compared to clean seawater, polluted seawater has the following characteristics:

Sulfide ions, dissolved oxygen level is low (5 ppm or less), chemical oxygen demand (COD) is high (COD is normally 2 to 4 ppm in clean seawater but tends to rise with pollution) and the pH is low. Clean seawater is 8.0 to 8.3, but polluted seawater drops under 8.

Stress corrosion cracking (SCC)- it occurs when a material under tensile stress is exposed to a corrosive environment, and for copper alloy tube it is well known that ammonia is the corrosion catalyst. Aluminum brass is susceptible to cracking in fresh water, so considerable care is required in usage, because stress concentrates on the corrosion pits, causing high localized stress loads.

Sand erosion refers to erosion after damage to the protective film from sand transported by the cooling seawater. It may take the form of longitudinal erosion scars.

Copper zinc alloy tubes like Admiralty Brass or Aluminum Brass tubes are susceptible to stress corrosion cracking due to presence of aqueous Ammonia in overhead condensers.

Cupro-Nickel alloy tubes in overhead condensers corrode when they are exposed to hydrocarbon vapors containing H2S. Sulphide scales of nickel and Copper are formed in alkaline medium.

Erosion of tube ends is common in exchangers and is more pronounced where hydrocarbon streams contain solid particles. This phenomenon can be seen in exchangers in FCC unit.

Grooving around the tubes may take place at baffle locations due to vibrations or crevice corrosion.

Erosion-corrosion occurs when the erosion effects of the coolant removes the protective film, thus exposing a fresh surface to corrosion. This type of attack occurs mainly at the tube ends. High velocity, abrupt change in flow direction, entrapped air and solid particles will promote erosion corrosion of tubes in coolers and condensers.

Tubes in exchangers and coolers are also susceptible to bulging or warping due to long exposure to high temperatures above design range and may finally result in cracking.

Continued vibrations caused due to high velocity or pulsating vapors striking the tubes may lead to fatigue cracks or corrosion fatigue in the form of circumferential fracture of the tubes.

When steam is used as a heating medium in tube side of exchangers and reboilers, the condensate may cause grooving or pitting in the tubes.

Cooler tubes are susceptible to overheating due to partial/total blocking caused either by Low velocity of water or suspended solids in cooling water and or water outlet high temperature resulting in the deposition of hard $CaCO_3$.

Tubes in coolers and condensers are prone to localized pitting, Dezincification or denickelification.

Dezincification removes zinc from the brass alloy, leaving behind a porous, copper-rich brittle structure. When this reaction occurs in a limited area it is usually called plug type dezincification and when over a large area layer type dezincification.

Dezincification is said to be likely to occur in the following cases:

1) When in contact with slightly acidic or alkali seawater, with low dissolved oxygen content.
2) When seawater velocity is excessively low.
3) When tube temperature is excessively high.
4) When tube surface is covered with penetrative sediments or film.

5.3 Degradation at Tubesheets

Non-ferrous tubesheets that are mainly used in heat exchangers, like Naval Brass or Cupro-Nickel are susceptible to dezincification or denickelification in cooling water service.

Where tubes are eroded or corroded, tubesheets may get damaged at the tube ligament areas by formation of rat holes.

Solid particles or marine growth that settle down on the tubesheets due to inadequate filtering or screening of cooling water will cause localized attack on tubesheets.

Galvanic corrosion of tubesheets may take place at pass partition grooves when partition plates of channel or floating head cover made of noble metal like Monel or stainless steel get in contact with tubesheets.

5.4 Degradation at Floating Head Cover

The floating head covers that are generally made of carbon steel or lined with corrosion resistant metals such as Monel or Lead, may get corroded in water service at bolt holes and the holes get enlarged.

The carbon steel pass partition plates, which are in contact with nonferrous tube sheets, undergo galvanic corrosion.

The floating head back up rings corrode due to retention of acidic condensate in overhead condensers.

Failure of gaskets may cause crevice corrosion on gasket face of floating head cover flange.

Low alloy strength steel stud bolts, for example ASTM-A-193 Gr. B7, crack due to sulphide stress cracking phenomenon in overhead condensers handling sour gases in presence of moisture.

5.5 Degradation at Channel and Channel Cover

Channel and channel covers are prone to water side corrosion in coolers and condensers. Carbon steel pass partition plates corrode by galvanic action if they get in contact with the noble metallurgy of tubesheets. The bare and unlined carbon steel channel covers are prone to pitting and tuberculation corrosion. Monel lined or lead lined channels get corroded at defects in lining or its welds.

5.6 Degradation at Baffles

The baffle plates are mainly carbon steel plates and get thinned out due to general condensate corrosion in hydrocarbon streams.

Baffle holes get enlarged due to erosion and tube vibration.

5.7 Gaskets and Gasket Seating Surfaces

It is recommended that when a heat exchanger is dismantled for any reason, a new gasket to be used for re-assembling. This will tend to prevent future leaks and / or damage to the seating surfaces of the heat exchanger. Composition gaskets dry out and become brittle so that they do not always provide an effective seal when reused. Metal or metal jacketed gaskets, when compressed initially, yield to match their contact surfaces and work hardened. If reused, may provide an imperfect seal or result in deformation and damage to the gasket contact surfaces of the exchanger.

Gasket seating surfaces of floating head cover, shell cover, channel & tube bundles are prone to scouring, erosion & pitting due to leakages from flanges.

Fig. 5.1 Shell/ Tube heat exchangers

CHAPTER-VI

INSPECTION OF HEAT EXCHANGERS

6.1 SAFETY

Safety precautions must be taken before any equipment is entered. In general, these precautions include but not limited to the isolating the energy sources, check and reduction of confined space temperatures, flow of air, removal of sludge, gases, chemicals and other stream materials before entering.

The presence of toxic residues inside the equipment must be reviewed. Inspectors and maintenance crew who enter the equipment must wear Protective equipment such as coverall, hard hat, gloves, ear protector, face mask, glass and respirator if necessary.

Before the inspector or other crew enters inside the confined space, a work permit must be obtained and Lock Out-Tag Out put in place as per plant regulations. A safety officer or operator must enter or examine first to check and confirm hazard free conditions exist and the unit is isolated. It is required any hazardous chemical are removed before any work permit is issued.

It is mandatory to have safety orientation for those inspectors and maintenance crew who are assigned by contractors and are new to the area and may not be aware of the hazards in the plant. Contractor shall be advised of their responsibilities on safety for all personnel, including their own staffs, property and environment as per company regulation or jurisdictional laws in place.

Plant owners must ensure all personnel including the visitors are trained and briefed adequately on the safe working and they all are familiar with the procedures and instructions in place.

Inspector should have the copies of all regulatory documents applicable together with client specific procedures to be followed at inspection site. It is highly recommended to review the history of inspections and previous repairs or modifications on the unit. The details of the material, design data and information to be reviewed and recorded in inspector's note- book for reference and reporting. One of the required check points before the inspection is the review and copying of name plate data to ensure all information do match.

It is always a good idea to have a checklist made before the inspection to ensure all sections are covered and no part is missed out.

6.2 CLEANING

Heat exchanger shell, channel and tube bundle must be inspected before washing to determine any problems, including poor circulation, tubes blocking and broken parts.

After the initial inspection and taking samples, as required, the tube bundle may be removed, if possible, for cleaning and detailed inspection.

Inside of shell and tubes should be washed down thoroughly to remove mud, loose scale or similar deposits.

6.3 INSPECTION

It is the owner responsibility to ensure that an asset integrity management system is in place and strictly adhered to. Inspection plan for each heat exchanger must be prepared and available. The plan may be either risk based or time based. The inspection plan shall address the time and type of inspection for each unit.

Each heat exchanger shall have the critical points with respect to internal corrosion or other failure mechanism identified and clearly marked for inspection or examination during on-stream inspection.

It must be reminded that this book focusses on intrusive and non-intrusive inspection of units that are installed and commissioned.

6.3.1 On-Stream Inspection

The thickness reading is normally performed every year after the installation for three to five years to establish the corrosion rate. Integrity engineers may decide on the frequency of the measurement afterward.

Each heat exchanger shall be inspected visually and thoroughly for any sign of damage, corrosion, leakage or vibration when the item is in use without disturbing the operation.

The purpose of this inspection is to ensure that the item is working properly and to collect as much information as possible to assess the integrity of the unit. The visual inspection is the basic and very important part that ensures the following:

✓ The foundation is without any crack or settlement
✓ No leak sign exists in any connection
✓ No vibration or abnormal noise exist
✓ The coating in sound and no sign of corrosion or rust exists
✓ Thermal insulation, if any, is intact with no damage
✓ No deformation exists on the body of the heat exchanger
✓ All expansion joints and pipe support or hangers are in place
✓ Name plate is in place and visible
✓ Thickness Measurement locations and ports (in case of insulated unit) are clearly marked and numbered.
✓ Earthing connection is in place and connected.

✓ Conditions of ladders, platforms, connections, fittings are satisfactory.
✓ It is necessary to open inspection windows on insulated exchangers at locations vulnerable to corrosion (CUI) or deterioration.

Experienced inspectors will use checklist to record the result of the visual inspection with notes. The checklist will help inspector not to miss any part.

The complementary part of in-service inspection is thickness measurement of the heat exchanger body and connections. It is performed to record the remaining thickness and evaluate the corrosion rate. All the reading must be made by a trained technician to ensure the instrument is calibrated and the measured thicknesses are accurate.

6.3.2   Internal Inspection

This type of inspection is intrusive and takes place with planned or unexpected shut down of the unit and provides better chance to access the internal parts.

A thorough visual inspection is the most powerful tool in a heat exchanger integrity management program. The examination is not limited to exchanger components. The fluid is monitored with the details of flow, temperature, operating history, material performance, and other information.

Borescope is essential to examine the inside of heads and tubes. Some video probe systems utilizing a small camera attached to a flexible probe tip have been used for internal inspection of the tubes. In recent years, IRIS (Internal Rotary Inspection system) has been used which travels inside the tube with flexibility and capability of rotation and sends the video image to the monitor and can be recorded in digital media as a permanent record. The flexible nature of the IRIS allows for inspection of remote areas and U-bends. This type of inspection identifies pitting, corrosion, corrosion by-products, fouling and debris and is a strong tool for visual inspection of the tubes internal.

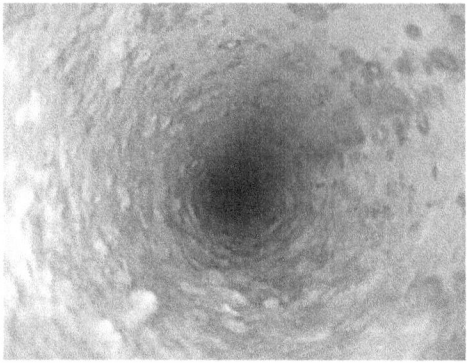

Fig.6.1- A picture of tube internal surface

The internal inspection mostly involves the removal of tube bundle and channels. A thorough visual inspection of all surfaces required to identify the corrosion, scaling, scratches, erosion and weld wash away. Any deficiency shall be photographed. Inspector can ask for NDT examinations if in doubt.

The tube bundle must be cleaned to remove the slug and trapped products in the space between the tubes. Hot water or steam is normally used for cleaning. This is also required to improve the fluid flow and heat transfer.

To ensure that the tubes can hold the internal and external pressure, each tube may be pressure tested at a pressure approved by engineer using water or other non-corrosive medium. The tubes that show thickness loss or leak can be plugged at both ends.

In fixed tubesheet designs, shell side inspection is not possible due to the limited access, however Borescope can be used to inspect the internal surface of the shell as much as possible. The video can also be kept in records.

A thorough visual examination of shell must include major welds. All welds that have been previously repaired, especially the ones repaired multiple times, must be checked. Weld patches in the shell or heads deserve attention as possible corrosion points.

Cladding and pass partition welds are examined with suitable NDT methods for soundness.

The welds in shell saddles and supports are examined for cracks and corrosion from exposure to the surrounding environment, which can be more corrosive than the fluids the exchanger carries. Likewise, the supports should have expansion provisions such as slots in the base bolt holes; check that the free end can slide to accommodate thermal expansion and contraction.

All nozzle attachments at the heads and the shell should be inspected. The bending or twisting of a nozzle can be a sign of excessive loading. Tangential nozzles must be checked carefully because of the inherent difficulty in their weld fit-up and attachment.

If any part of the shell or welds need repair or build up, the procedure must be prepared and approved by authorized inspector or asset integrity engineer and the work shall be monitored by the inspector. The repair of ASME U-Stamp units must be monitored by authorized inspector and all repairs be recorded. NB-23 requirements shall be adhered to.

Attention must be given to the seals in floating head units of the TEMA - P and W types. These seals must be disassembled and maintained per the manufacturer's instructions. Excessive tightening can cause extra load on the joint and restricts the tube-sheet movement, thus defeating the joint's purpose. Because of this, a small amount of packing fluid leakage is acceptable. The metal sealing surfaces ought to be protected carefully when they are handled because small dents, scratches and scrapes will damage the sealing ability. A small defect when applied to a high-pressure fluid containment is serious because fluid passing through the defect at high velocity quickly damages the adjacent metal surfaces.

Fatigue cracks in the tubesheet ligaments can signify excessive tube vibration that is transmitted to the tubesheet via the tube-to-tubesheet joint. It is critical to identify the basis for tube-to-tubesheet joint failure (erosion, corrosion, vibration and others). A joint can be repaired only a few times before the material is work-hardened or over-stressed.

The tie rods, baffles, and spacers are examined for damage, particularly for corrosion from the fluid or from chemical cleaners, if used. Failure of these components can shift and/or rotate the tube bundle and severely reduce the exchanger capacity. The gaps between tie rods, baffles, and spacers are good candidates for fouling and corrosion.

CHAPTER VII

CODES AND STANDARDS

7.0 CODES AND STANDARDS

7.1 American Standards

American Society of Mechanical Engineers (ASME) is the leading authority for publishing pressure vessel code. ASME section-1 is on Power Boiler and is used for the design, construction, inspection and stamping of power boilers. There are many manufacturers worldwide that have authorization from ASME to apply code stamp on power boilers.

ASME Sec.1   Power Boiler

Section IV   Rules for construction of
                Heating Boilers
Section VI   Recommended Rules for

Care and Operation of
Heating Boilers

Section VII   Recommended guidelines
for the care of power boilers

ASME QAI-1   Qualification for authorized
inspection.

Section VIII   Pressure Vessels, Divisions 1
and 2c (rules for construction
of pressure vessels including
deaerators, blow-off
separators, softeners, etc.)

Section IX:

Welding and Brazing Qualifications
(the section of the Code that defines
the requirements for certified
welders and welding.)

B-31.1   Power Piping Code

CSD-1 – Controls and Safety Devices for
Automatically Fired Boilers (applies to boilers with
fuel input in the range of 400 thousand and less
than 12.5 million Btuh input)

National Board Inspection Code. NB-23

API 510- Pressure Vessel inspection code

API573- Inspection of fired boilers and Heaters

API 660- Shell and Tube Heat Exchangers

7.2 Canadian Standards:

CSA B51   Boiler, pressure vessel and pressure piping code.

Alberta Pressure Equipment and Safety Authority (ABSA):

AB 525   Overpressure protection requirement for pressure equipment and pressure piping

AB 513   Pressure Equipment repair and alteration requirements

AB 505   Risk based inspection requirement for pressure equipment.

AB 506   Inspection and Servicing Requirements for In-Service Pressure Equipment

AB 526   In-Service Pressure Equipment
          Inspector Certification
          Requirements

7.3 British Standards:

BS EN 12952-1: 2001
Water-tube boilers and auxiliary installations-
General

BS EN 12952-2: 2011
Water-tube boilers and
auxiliary installations- Materials for
pressure parts of boilers and accessories.

BS EN 12952-3:2011
Water-tube boilers and auxiliary installations,
Design and calculation of pressure parts.

BS EN 12952-5: 2011
Water-tube boilers and auxiliary installations,
workmanship and construction of pressure parts
of the boiler.

BS EN 12952-7:2011
Water-tube boilers and auxiliary installations,
inspection during construction, documentation
and marking of pressure parts of the Boilers

BS EN 12952-10:2002

Water-tube boilers and auxiliary installations, requirements for safeguards against excessive pressure.

BS EN 12952-11: 2007
Water-tube boilers and auxiliary installations requirements for limiting devices of the boiler and accessories.

BS EN 12953-1: 2002   Shell boilers- General

BS EN 12953-2: 2012
Shell boilers- Materials for pressure parts of boilers and accessories.

BS EN 12953-5:2002
Shell boilers- Inspection during construction documentation and marking of pressure parts of the boilers

BS EN 12953-6:2002
Shell boilers- Requirements for equipment for the boilers.

BS EN ISO 16812-2019
Shell and Tube Heat Exchangers

7.4   European Standard

EN 12952
Standard for water-tube boilers
EN 12953
Standard for shell boilers

7.5 German Standards:

DIN 2918
Stationary shell boilers of welded construction (other than water-tube boilers)

7.6 French Standards:

NFE 31-001   Boilers operating with solid,
                      Liquid, or gas fuels
NFE 32-100   National Standard for boilers

7.7 International Standard Organization (ISO)

ISO 831
Rules for construction of stationary boilers

ISO 570
Stationary shell boilers of welded construction (other than water-tube boilers)

7.8 Russian Standard

GOST 31842-2012 Petroleum and natural gas industries. Shell-and-tube heat exchangers. Technical requirements

GOST 34233.7-2017 Vessels and apparatus. Norms and methods of strength calculation. Heat exchangers

GOST R ISO 15547-1-2009 Petroleum and natural gas industries - Plate-type heat exchangers - Part 1: Plate-and-frame heat exchangers

GOST R 52630 (GOST 34347-2017) Steel welded vessels and apparatus. General specifications

GOST 3619 Stationary steam boilers. Types and basic parameters

RD 10-249-98 Norms of strength calculation of stationary boilers and pipelines of steam and hot water.

References

1- C.B. Oland for OaK Ridge National Laboratory, 2002- Guide to low emission boiler and combustion equipment selection.
2- Mohammad A. Malek, Ph.., PE- Power Boiler Design, Inspection and Repair. McGraw-Hill ,2004
3- Alec Groysman, Corrosion problems and solutions in oil refining and petrochemical industry, springer.
4- Elayaperumal, Y. S. Roy-Corrosion failures, theory, case studies and solutions.,2015
5- K.R.Rao- Companion guide to ASME -Boiler and pressure vessel codes Vol.1, 5th ED. 2018
6- V.Ganapathy, ABCO industries- Industrial boilers and heat recovery steam generators, 2003
7- Kenneth E. Heselton,PE,CEM- Boiler operator's handbook, 2005
8- Clifford Mathews- Engineer's guide tp pressure equipment, 2001
9- Suez Water technologies & Solutions- Handbook of Industrial solutions.

AUTHORS

Fereydoun Majdnia

Fereydoun was born, brought up and educated in Iran. His initial study was pure mathematics in Tabriz University and then mechanical engineering at Abadan Institute of Technology. He worked 15 years in offshore oil company in Iran. He then moved abroad and worked with ABS Group an inspection and certification company as senior inspector and later he joined Technip Middle East an EPC company as quality manager. He has worked in Dubai, Abu Dhabi, Kuwait, Canada and Moscow.

Evgeny Sergeev

Evgeny is born and educated in Russia. He studied at Tomsk University and graduated on control. He pursued inspection and got series of training on quality system and various standards. Evgeny was involved in automotive industry on localization of the parts and supplier. He then joined NIPPIGAS an EPC company under SIBUR, the famous and huge petrochemical holding company in Russia as quality system manager where he currently works and looks after the quality of the delivered services by international contractors.